REGENERATIVE Soil

the teacher's guide

written by
Matt Powers

Copyright © 2020 Matt Powers
All Rights Reserved
Written by Matt Powers
Illustrations by Matt Powers and at least one by Brandon Carpenter as labeled
Cover is a combination of photography by Matt Powers
Charts, Illustrations, and diagrams by Matt Powers unless labeled otherwise
Most photos are from Wikimedia, Unsplash, and other open source outlets.
Other photography by Matt Powers unless labeled otherwise–all photos are used with author permission
Formatted by Matt Powers
Printed on 100% recycled paper in China

Send all Inquiries to:
Matt Powers
Matt@ThePermacultureStudent.com

Published and Distributed by PowersPermaculture123
ISBN: 978-1-953005-99-1
website: www.RegenerativeSoilScience.com

Please Note: Just as food that nourishes one person causes an allergic reaction in another, the same concept of situational complexity applies to soils, water, microbes, contaminants, and more that is situationally complex. In this book, complexity is embraced with the understanding that every situation and biome is unique. The information in this book represents research from sources listed–it is an educational and informational resource and does not represent any agreement, guarantee, or promise by any party associated with the creation or editing of this book. The publisher, editors, and author are not responsible for any negative or unintended consequences from applying or misapplying any of the information in this book.

TABLE OF CONTENTS

Introduction (1)

Guiding Educational Principles & Concepts (3)

Standards Used & Content Rationale (9)

Next Generation Science Standards (11)

National Science Education Standards (14)

Permaculture Education Standards (16)

1. What Is Soil? (17)

2. The History of Soil (20)

3. The History of Humans and Soil (22)

4. The Mineral Components of the Soil (25)

5. The Biological Components of the Soil (29)

6. The Actions (33)

7. Making Your Plan (36)

Projects (39)

Final Thoughts (54)

About the Author (55)

INTRODUCTION

Thank you for teaching others about soil! It's a lynchpin we often take for granted. You are investing in the future and in a better world: THANK YOU!

This is book is a teacher's guide companion to the book *Regenerative Soil* and the online course *Regenerative Soil*. Use it to teach workshops, high school electives (aligned to standards), in college settings, and through adult education. If you don't have to align to the standards, don't worry about those portions. If you are running an ungraded program, assess their learning in observational, survey, and novel ways to keep things light but also to verify that they are understanding and applying their learning.

There are suggested readings, courses, youtube videos, and books, but you don't have to have everything — many resources can be found synopsized (often by the authors themselves) online through free channels. The literature for most of full program is freely available online though all of these options would be greatly enriched by the paid programs and even subscriptions to journals like *Nature*. I include it all so you can pick and choose, plot your course, and set the timeframes for your sections yourself. When I teach this material in my online courses, I always am doing it in multiple sections like pt 1 and 2, etc. because some of these sections can go so deep — some can even turn into history/

literature courses on their own (like tracing David Montgomery's DIRT timeline would be an epic college course!) Setup like a choose-your-own-adventure manual, this book allows you to create a program that fits your style, direction, and audience.

GUIDING EDUCATIONAL PRINCIPLES & CONCEPTS

Did you know there is a succession to thought? It takes more understanding and experience to create something than it does to recognize or define it. If you cannot whistle but are old enough to understand whistling and recognize it (identification), you know that it is blowing air through your lips in a very precise manner (comprehension). You struggle to apply it until you have a breakthrough (application) and you begin to analyze HOW you just did that and how it all works (analysis), where are the limitations of your skill, and then you try to synthesize your new skill with a novel setting: whistling a song you know (synthesis). As you listen to yourself perform a desired note or an entire song, the critical mind and faculty awaken and you ask: was that good? How did that sound? What can I do better or differently the next note or run-through (critical thinking)? Lastly students can chart their own course and create their own unique expression of that concept or using that skill (creation). Almost everywhere educators are trained, they learn about Bloom's Taxonomy of

Bloom's Taxonomy of Cognition

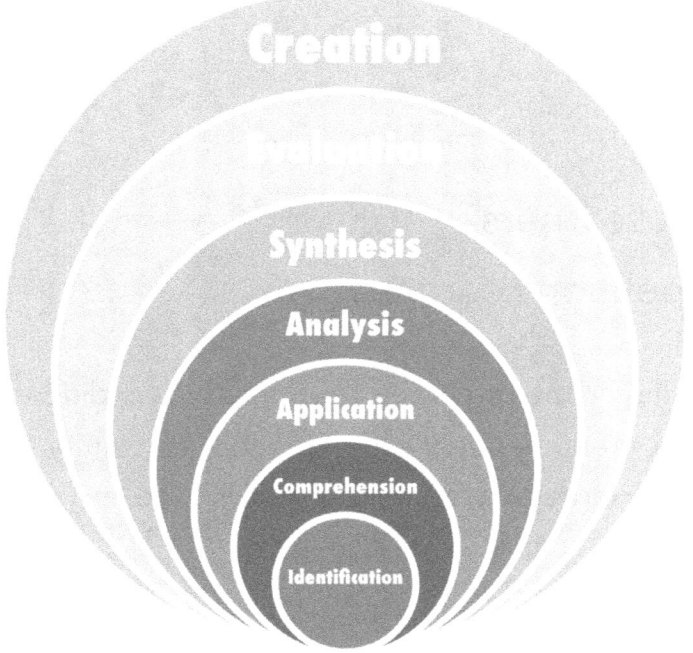

Cognition, so they can order their lessons to build in complexity in a logical succession of thought. It follows the way learners' thinking progresses.

Did you know that everyone learns differently… in different contexts? You likely know folks that can read and retain information differently than when they are told verbally, and vice versa. Some students learn through movement, some through listening, some through seeing it done, and some from using their hands while they are watching you doing it – it all depends. If you look up learning modalities, you typically just find: auditory, kinesthetic, and visual, but there are more. People are more nuanced, and we all have varying degrees of ability within each of these modalities. In addition, we can grow and improve in each and in all as well. As educators, we can use these modalities to reach more students, to stretch them further, and to connect their prior learning schema to new learning and even to transfer skills – as is what happens when an avid audiobook listener who is also learning to read transfers their listening comprehension skills to reading comprehension skills which at the same time leads to rapid improvements in reading retention, reading speeds, spelling, grammar, and writing skills.

Learning Modalities

- Auditory Learners
- Kinesthetic & Tactile Learners
- Visual & Spatial Learners
- Analytical Learners
- Holistic Learners
- Social & Verbal Learners
- Game & Competition Learners

Not only do we not learn in the same way, but we also have different intelligences, and these, too, can be strengthened through practice and study, but many of these intelligences are deeply ingrained and part of who we are as a person, thinker, and learner. This also forms subtle biases in us as educators – we praise and reward those who do well in our classrooms without recognizing that most succeed at a high level because they are already strong in our own particular form of intelligence. While we want those students to reach as high as possible in their own intelligence and modality, the students who are not already primed in that form of intelligence may be working harder to do satisfactory work in the same context. Every learner has their own unique context, prior learning, and strengths in a combination of intelligences though usually we are strong in one or two areas. Traditional schooling has focused on math/logical and verbal/linguistic for standardized testing and grading while ignoring the many other types of ways we learn, think, and express ourselves.

9 Forms of Intelligence

- Nature
- Music
- Body/Kinesthetic
- Spatial/Visual
- Interpersonal (communication between people)
- Intrapersonal (metacognition, understanding thought)
- Math/Logical
- Verbal/Linguistic
- Spiritual/Philosophical/Existential

Permaculture Education Principles

- We can always learn something new from carefully observing natural systems
- We all can learn to be regenerative in all ways though it will require time, effort, and adaptation

- Learning cannot be forced, only shared and accepted. No one can force anyone to learn anything. We can only invite, present, and offer a pathway. All learning is acceptance of a truth and that is always a choice
- Everyone can learn, grow, and progress though it requires a growth mind-set to do so
- "Kids do well when they can" – Dr. Ross Greene. Students fail when they lack the skills, schema, or confidence (which comes from competence)
- Our problems can become our greatest solutions and can transform us if we have a growth mindset
- Teaching is the highest level of learning – have your students teach as much as possible
- Enthusiasm is the currency of a vibrant learning community, like sunlight to a forest – <u>Bring the Enthusiasm</u>
- Beyond Socratic is Heuristic – help students ask better questions by having them practice more often
- Reflection is key to deeper understanding, problem solving, and adaptation
- Authentic Permaculture Education leads to Regenerative Action & Empowerment
- Holistic empowerment is key to maintaining sustained growth
- Pairing regenerative practices with service to the community and those in need deeply encodes the learning and significance of that learning
- Self care is central to all care

Best Practices for Permaculture Educators

- Reach your students through their own modalities, prior knowledge, and experience
- Build trust in your classroom and community – it is the basis for the feeling of safety and comfort needed for deep learning
- Regularly connect new learning to permaculture principles and the 3 Ethics, especially when the focus becomes more technical, complex, or hands-on
- Use frameworks to organize thinking and designs, but be flexible and be ready to adapt, combine, and rethink
- Create lessons and projects with choices, like a menu, and present them as an invitation

- Create lessons and projects that rely upon, strengthen, and support all forms of intelligence not just math and verbal
- Allow students to express their learning in modalities they are comfortable in, but challenge your students to challenge themselves
- Use their holistic goals as the foundation for self-appointed challenges based in curiosity
- Everyone has a story – facilitate an environment and experience where all can feel they can share their own story
- Keep lessons, projects, and themes student-centered to maintain engagement and deep learning
- Use projects to bundle skills and learning into authentic learning experiences and performance assessments
- Invite students to take on the role of the teacher as a summative assessment of their learning
- Use groups to support peer-to-peer learning and sharing
- Model best practices for self-improvement and self-care: positive self-talk, stepping out problems with analysis and critical thinking, regular meditation, regular exercise, and healthy spiritual and social relationships. This does matter – when we present holistic health in an accessible, generalized manner, we help students find that balance as well even when it isn't modeled at home and even when it is nothing like what they believe or will do to maintain their holistic health
- Model and use nonviolent communication especially with students in conflict or crisis situations
- Use restorative justice practices like restorative circles to mediate conflict and to resolve conflicts
- Build a classroom community of growth mind-set life learners who see mistakes as part of the learning process and critical to growth
- Meet your students where they are at and at their level – this may mean at their eye level, sitting among the students instead of off to the side, or organizing the lesson around their interests and zone of proximal development
- Foster a community of lateral, cooperative, peer-to-peer learning and decision making by using questions and prompting students to take the lead asking questions and planning projects

- Build a community wherein social permaculture principles and nonviolent communication are practiced habit

STANDARDS USED & CONTENT RATIONALE

Working with soil is intrinsically a scientific act — it is such a foreign territory that we MUST rely upon the scientific method of investigation and analysis to even comprehend what we are seeing, doing, and influencing. It covers and touches upon so many disciplines that we normally separate in scientific study: biology, microbiology, geology, chemistry, biochemistry, geochemistry, geomorphology, physics, biophysics, environmental science, soil science, agriculture, horticulture, bioremediation, and even to a degree entymology. It's also hands-on and goals-oriented, so it lends itself perfectly to the current science education paradigms of inquiry, observation, quantification, analysis, and scientific solution creation to problems ranging from the pragmatic to the global.

Working with soil is also the best way to learn chemistry and prepare for or enrich any agricultural, horticultural, FFA, gardening, grazing, environmental science, or landscaping program. One of the best reasons it is perfect for learning chemistry is it is hands-on, ties pH to Eh in a fundamental way that changes one's perspective on all life and natural systems, and deeply explores the behaviors and roles that essential elements play in those systems. By working with soil, we learn to recognize and express what we see and observe with technology in terms of chemistry through the interactions of life (biochemistry), and this is the best context to be introduced to elements and chemistry because it is so messy. We see both chemistry in action and accept the complexity of interaction and influences that are at play in the soil — it is a perfectly humble mindset grounded in a simple fluency that will allow the learner to grow, learn, and apply their learning with care, precision, and reflection.

Everyone should know how to work with soil. Our world revolves around soil as it is the medium for life. It is our job as members of this planet to be stewards and creators of regenerative soil. We can solve so many of our problems with soil, and as you and your students will learn, many of these soil microbes and solutions are in widespread use in specific industries, areas, and practices but not yet fully understood or adopted widely. *Regenerative Soil* brings all the solutions together, all the pieces of the jigsaw solution to our soil crisis (and in tandem, the solutions to the climatic, agricultural, hunger, and nutritional crises). The solutions not only provide pathways to lead us away from these crises, but they open up new avenues for entrepreneurship, innovation, and job creation. Top quality soil has never been in such high demand. Farmers, orchardists, and ranchers the world over are seeking solutions to the ever greater pest, drought, and deficiency pressures they are almost all experiencing. This program offers a pathway to real opportunity and service for all.

This program is also superb for making garden management a measurable and educational endeavor. Many schools install gardens only to see them neglected and their care not being tied to curriculum standards. By making soil understandable from so many avenues, we can connect and overlap this program with all core science education and almost all electives. It will enrich all other programs as it makes their concepts real and tangible to the learner.

Thank you for teaching the next generation about REGENERATIVE SOIL!!

NEXT GENERATION SCIENCE STANDARDS
(NGSS STANDARDS)

HS-LS2 Ecosystems: Interactions, Energy, and Dynamics

HS-LS2-3. Construct and revise an explanation based on evidence for the cycling of matter and flow of energy in aerobic and anaerobic conditions.

HS-LS2-5. Develop a model to illustrate the role of photosynthesis and cellular respiration in the cycling of carbon among the biosphere, atmosphere, hydrosphere, and geosphere.

HS-LS2-6. Evaluate claims, evidence, and reasoning that the complex interactions in ecosystems maintain relatively consistent numbers and types of organisms in stable conditions, but changing conditions may result in a new ecosystem.

HS-LS2-7. Design, evaluate, and refine a solution for reducing the impacts of human activities on the environment and biodiversity.

LS2.A: Interdependent Relationships in Ecosystems

LS2.B: Cycles of Matter and Energy Transfer in Ecosystems

LS2.C: Ecosystem Dynamics, Functioning, and Resilience

LS4.D: Biodiversity and Humans

PS3.D: Energy in Chemical Processes

ETS1.B: Developing Possible Solutions

HS-ESS2 Earth's Systems

HS-ESS2-2. Analyze geoscience data to make the claim that one change to Earth's surface can create feedbacks that cause changes to other Earth systems.

HS-ESS2-4. Use a model to describe how variations in the flow of energy into and out of Earth's systems result in changes in climate.

HS-ESS2-5. Plan and conduct an investigation of the properties of water and its effects on Earth materials and surface processes.

HS-ESS2-6. Develop a quantitative model to describe the cycling of carbon among the hydrosphere, atmosphere, geosphere, and biosphere.

HS-ESS2-7. Construct an argument based on evidence about the simultaneous coevolution of Earth's systems and life on Earth.

ESS2.A: Earth Materials and Systems

ESS2.C: The Roles of Water in Earth's Surface Processes

ESS2.D: Weather and Climate

ESS2.E: Biogeology

HS-ESS3 Earth and Human Activity

HS-ESS3-2. Evaluate competing design solutions for developing, managing, and utilizing energy and mineral resources based on cost-benefit ratios.

HS-ESS3-4. Evaluate or refine a technological solution that reduces impacts of human activities on natural systems.

HS-ESS3-6. Use a computational representation to illustrate the relationships among Earth systems and how those relationships are being modified due to human activity.

ESS3.A: Natural Resources

ESS3.C: Human Impacts on Earth Systems

ESS3.D: Global Climate Change

ETS1.B: Developing Possible Solutions

HS.Matter and Energy in Organisms and Ecosystems

HS-LS1-5. Use a model to illustrate how photosynthesis transforms light energy into stored chemical energy.

HS-LS1-6. Construct and revise an explanation based on evidence for how carbon, hydrogen, and oxygen from sugar molecules may combine with other elements to form amino acids and/or other large carbon-based molecules.

HS-LS2-3. Construct and revise an explanation based on evidence for the cycling of matter and flow of energy in aerobic and anaerobic conditions.

HS-LS2-5. Develop a model to illustrate the role of photosynthesis and cellular respiration in the cycling of carbon among the biosphere, atmosphere, hydrosphere, and geosphere.

LS1.C: Organization for Matter and Energy Flow in Organisms

HS.Human Sustainability

HS-ESS3-1. Construct an explanation based on evidence for how the availability of natural resources, occurrence of natural hazards, and changes in climate have influenced human activity.

NATIONAL SCIENCE EDUCATION STANDARDS
(NSES GRADES 9-12 CONTENT STANDARDS)

CONTENT STANDARD A
As a result of activities in grades 9-12, all students should develop:
- abilities necessary to do scientific inquiry
- understandings about scientific inquiry

CONTENT STANDARD B
As a result of their activities in grades 9-12, all students should develop an understanding of:
- structure of atoms
- structure and properties of matter
- chemical reactions
- motions and forces
- conservation of energy and increase in disorder
- interactions of energy and matter

CONTENT STANDARD C
As a result of their activities in grades 9-12, all students should develop understanding of:
- the cell
- molecular basis of heredity
- biological evolution
- interdependence of organisms
- matter, energy, and organization in living systems
- behavior of organisms

CONTENT STANDARD D
As a result of their activities in grades 9-12, all students should develop an understanding of:
- energy in the earth system
- geochemical cycles
- origin and evolution of the
- earth system
- origin and evolution of the universe

CONTENT STANDARD F

As a result of activities in grades 9-12, all students should develop understanding of:
- personal and community health
- population growth
- natural resources
- environmental quality
- natural and human-induced hazards
- science and technology in local, national, and global challenges

CONTENT STANDARD G

As a result of activities in grades 9-12, all students should develop understanding of:
- science as a human endeavor
- nature of scientific knowledge
- historical perspectives

PERMACULTURE EDUCATION STANDARDS
(PES GRADES 9-12 CONTENT STANDARDS)

2. Knowledge

2.5 Students are able to identify, create detailed representations of, and teach the water cycle, the mineral cycle, the carbon cycle, and the global annual seasonal cycle in relation to the sun especially in relation to climate change and desertification

2.6 Students are able to identify, describe, and present the different components of soil, the soil food web, photosynthesis in relation to the soil food web, and ways to improve soil and soil food web interactions

2.7 Students are able to identify, describe, illustrate, and present the various interactions
and function of trees and forests in relation ecosystems, natural cycles, precipitation, watersheds, soil, fungi, climate change, wildfires, and all biodiversity

3. Design

3.2 Students are able to create an advanced permaculture design for a home site – one that addresses water, waste, food, soil building, energy, and shelter

3.3 Students are able to design, set up, and manage a small outdoor garden, indoor garden, greenhouse garden, and larger outdoor garden

4. Regenerative Skills

4.4 Students are able to make and apply thermophilic (hot) compost and vermicompost (using worms) to process food waste as well as compost teas and extracts

4.5 Students are able to use a microscope to identify soil food web organisms and determine soil, compost, compost extract, and compost tea quality

4.14 Students have participated in regular acts of large-scale land restoration

1. WHAT IS SOIL?

NGSS

LS2.A: Interdependent Relationships in Ecosystems
LS2.B: Cycles of Matter and Energy Transfer in Ecosystems
LS2.C: Ecosystem Dynamics, Functioning, and Resilience
ESS2.A: Earth Materials and Systems
ESS2.C: The Roles of Water in Earth's Surface Processes
ESS2.D: Weather and Climate
ESS2.E: Biogeology
ESS3.A: Natural Resources

HS-LS1-5. Use a model to illustrate how photosynthesis transforms light energy into stored chemical energy.

NSES

CONTENT STANDARD A – G

PES

2.7 Students are able to identify, describe, illustrate, and present the various interactions
and function of trees and forests in relation ecosystems, natural cycles, precipitation, watersheds, soil, fungi, climate change, wildfires, and all biodiversity

RATIONALE

It is necessary for students to learn the primary components of soil, their qualities and characteristics, their interactions, and their distribution of diversity across climates and bioregions before they explore the topic in full depth. This serves as an overview, foundational learning experience, and a process of correcting incorrect prior learning and assumptions.

OBJECTIVES

Students will be able to…
- Define and describe the components of soil and their characteristics, qualities, interactions, and distribution across climates and bioregions
- Describe soil and its components in a holistic, relatable, scientifically accurate, and clear fashion
- Define and describe the states and roles of water in relation to soil including exclusion zone water

- Define and describe the roles of air in relation to soil

LESSON CONTENT
- Lecture/Presentation/Video
- Reading/Listening/Watching
- Group Discussion
- Nature Immersion & Observation of Natural Systems
- Tactile Interaction
- Reflection
- Group Challenges & Activities

LESSON ACTIVITIES
- Read & Discuss *Regenerative Soil* sections: "Why Now?" and "The Nature of Soil", p. 1–16
- Watch & Discuss *Regenerative Soil, the Online Course* sections: "Course Introduction & Overview" and "The Nature of Soil"
- *Jar Soil Test* – Students will perform the Jar Soil Test from p. 140 of *Regenerative Soil*
- *Soil Texture Test* – Students will perform the Soil Texture Tests using their hands from p. 140–141 in *Regenerative Soil*
- *Observe & Analyze* – Students will observe the natural landscape and identify 3-5 distinct soil types, they will collect samples of each soil type, and they will perform Jar Soil Tests for each soil type, comparing them to each other

ASSESSMENTS
- Group Discussion (observed and through direct interaction)
- Group Activities, Artistic Representations, and Projects
- Written and Verbal Quizzes
- Identification Games/Exercises
- Student presenting (reteaching) content
- Student Notes
- Reflective Writing on New Learning

DIFFERENTIATIONS

Audiobook, Video-based instruction, teaching outside, having soil within reach for tactile interaction, with background music or in silence, discussion-based instruction, and learning 1:1, in small groups, or large groups.

2. THE HISTORY OF SOIL

NGSS

LS2.A: Interdependent Relationships in Ecosystems

LS2.B: Cycles of Matter and Energy Transfer in Ecosystems

LS2.C: Ecosystem Dynamics, Functioning, and Resilience

HS-ESS2-5. Plan and conduct an investigation of the properties of water and its effects on Earth materials and surface processes.

HS-ESS2-7. Construct an argument based on evidence about the simultaneous coevolution of Earth's systems and life on Earth.

ESS2.A: Earth Materials and Systems

ESS2.C: The Roles of Water in Earth's Surface Processes

ESS2.D: Weather and Climate

ESS2.E: Biogeology

ESS3.A: Natural Resources

NSES

CONTENT STANDARD A – G

PES

2.7 Students are able to identify, describe, illustrate, and present the various interactions
and function of trees and forests in relation ecosystems, natural cycles, precipitation, watersheds, soil, fungi, climate change, wildfires, and all biodiversity

RATIONALE

In order to understand soil's unique relationship within all ecosystems and with all life, students need to know the history of soil creation in relation to natural forces and to the evolution of life on earth.

OBJECTIVES

Students will be able to…
- describe how soil was created over time through natural processes (geomorphology)
- describe the forces at work on soil
- describe how soil creation and the evolution of life are interconnected

LESSON CONTENT
- Lecture/Presentation/Video
- Reading/Listening/Watching
- Group Discussion
- Nature Immersion & Observation of Natural Systems
- Tactile Interaction
- Reflection
- Group Challenges & Activities

LESSON ACTIVITIES
- Review & Discuss *Regenerative Soil* sections: "The Nature of Soil", p. 8–16
- Review & Discuss *Regenerative Soil, the Online Course* section: "The Nature of Soil"
- *Observe Natural Phenomenon* – Students will visit areas of natural phenomenon where soils are being eroded, formed, or transported, describe them, take photographs, and document the soil in transformation.
- *Soil Erosion Comparison* – Students will acquire 3 different soil types and run water passively through them to measure water retention capacities of different soils in relation to their organic matter and clay levels.

ASSESSMENTS
- Group Discussion (observed and through direct interaction)
- Group Activities, Artistic Representations, and Projects
- Written and Verbal Quizzes
- Identification Games/Exercises
- Student presenting (reteaching) content
- Student Notes
- Reflective Writing on New Learning

DIFFERENTIATIONS
Audiobook, Video-based instruction, teaching outside, having soil within reach for tactile interaction, with background music or in silence, discussion-based instruction, and learning 1:1, in small groups, or large groups.

3. THE HISTORY OF HUMANS & SOIL

NGSS

LS2.A: Interdependent Relationships in Ecosystems
LS2.B: Cycles of Matter and Energy Transfer in Ecosystems
LS2.C: Ecosystem Dynamics, Functioning, and Resilience
LS4.D: Biodiversity and Humans
PS3.D: Energy in Chemical Processes
HS-ESS2-2. Analyze geoscience data to make the claim that one change to Earth's surface can create feedbacks that cause changes to other Earth systems.
HS-ESS2-4. Use a model to describe how variations in the flow of energy into and out of Earth's systems result in changes in climate.
ESS2.A: Earth Materials and Systems
ESS2.C: The Roles of Water in Earth's Surface Processes
ESS2.D: Weather and Climate
ESS2.E: Biogeology
HS-ESS3-6. Use a computational representation to illustrate the relationships among Earth systems and how those relationships are being modified due to human activity.

ESS3.A: Natural Resources
ESS3.C: Human Impacts on Earth Systems
ESS3.D: Global Climate Change
HS-ESS3-1. Construct an explanation based on evidence for how the availability of natural resources, occurrence of natural hazards, and changes in climate have influenced human activity.

NSES

CONTENT STANDARD A – G

PES STANDARDS

2.5 Students are able to identify, create detailed representations of, and teach the water cycle, the mineral cycle, the carbon cycle, and the global annual seasonal cycle in relation to the sun especially in relation to climate change and desertification

2.7 Students are able to identify, describe, illustrate, and present the various interactions
and function of trees and forests in relation ecosystems, natural cycles, precipitation, watersheds, soil, fungi, climate change, wildfires, and all biodiversity

RATIONALE

Students need to understand the history of human interaction with soil, the mostly detrimental nature of this interaction, and the inherited effects that today's soils and bioregions face due to this interaction.

OBJECTIVES

Students will be able to...
- describe the process and factors of soil loss over time due to human activity
- describe the effects of soil loss on civilizations (present and historical)
- Describe how soil loss contributes to and leads to desertification, climate change, eutrophication, dead zones, and ocean acidification

LESSON CONTENT

- Lecture/Presentation/Video
- Reading/Listening/Watching
- Group Discussion
- Nature Immersion & Observation of Natural Systems
- Tactile Interaction
- Reflection
- Group Challenges & Activities

LESSON ACTIVITIES

- Read & Discuss *Regenerative Soil* sections: "Humans & Soil", p. 129–137
- Watch & Discuss *Regenerative Soil, the Online Course* section: "Humans & Soil"
- Read & Discuss *DIRT: Erosion of Civilizations* by David Montgomery
- Read & Discuss *Kiss the Ground* by Josh Tickell
- Watch & Discuss footage or a documentary on the America Dust Bowl
- Watch & Discuss *Symphony of the Soil*, the documentary
- Watch & Discuss *Kiss the Ground,* the documentary
- Write an essay or create a presentation/video on how manmade soil loss has negatively affected ecosystems and civilization itself
- Write an essay on how natural resources have influenced history
- Generate a diagram of the flow of energy in and out of the soil environment and its relationship to climate

- Generate a diagram (ideally on a computer) demonstrating human influences on soil

ASSESSMENTS

- Group Discussion (observed and through direct interaction)
- Group Activities, Artistic Representations, and Projects
- Written and Verbal Quizzes
- Identification Games/Exercises
- Student presenting (reteaching) content
- Student Notes
- Reflective Writing on New Learning

DIFFERENTIATIONS

Audiobook, Video-based instruction, teaching outside, having soil within reach for tactile interaction, with background music or in silence, discussion-based instruction, and learning 1:1, in small groups, or large groups.

4. THE MINERAL COMPONENTS OF SOIL

NGSS

HS-LS2-5. Develop a model to illustrate the role of photosynthesis and cellular respiration in the cycling of carbon among the biosphere, atmosphere, hydrosphere, and geosphere.
LS2.A: Interdependent Relationships in Ecosystems
LS2.B: Cycles of Matter and Energy Transfer in Ecosystems
LS2.C: Ecosystem Dynamics, Functioning, and Resilience
PS3.D: Energy in Chemical Processes
HS-ESS2-6. Develop a quantitative model to describe the cycling of carbon among the hydrosphere, atmosphere, geosphere, and biosphere.
ESS2.A: Earth Materials and Systems
ESS2.D: Weather and Climate
ESS2.E: Biogeology
ESS3.A: Natural Resources
ESS3.D: Global Climate Change

NSES

CONTENT STANDARD A – G

PES STANDARDS

2.5 Students are able to identify, create detailed representations of, and teach the water cycle, the mineral cycle, the carbon cycle, and the global annual seasonal cycle in relation to the sun especially in relation to climate change and desertification
2.7 Students are able to identify, describe, illustrate, and present the various interactions
and function of trees and forests in relation ecosystems, natural cycles, precipitation, watersheds, soil, fungi, climate change, wildfires, and all biodiversity

RATIONALE

Students need a holistic understanding of the essential soil minerals for plants in order to properly work with and comprehend soil: mineral roles, attributes, interactions, forms, states, and cycles. Students also need to be able to recognize and treat toxicity or deficiency of soil minerals. Students need to understand the way we analyze and compare different soils including pH, Eh, and paramagnetism.

OBJECTIVES

Students will be able to…

- List and describe the elements that are essential nutrients for all plants
- Define and describe the essential nutrient cycles, forms, states, roles, and interactions for each element
- Define and describe mineral antagonism in the soil and its effects on nutrient availability and deficiency
- Recognize and describe mineral toxicity and deficiency in plants
- Define and describe the pH scale and its effects, implications, influence, and distribution
- Define and describe the Eh scale and its effects, implications, influence, and distribution
- Define and describe the paramagnetism and its effects, implications, influence, and distribution
- Define and describe CEC including its effects, implications, and influence
- Define and describe how to alter the pH, Eh, CEC, and paramagnetism of soils
- Define and describe adsorption and absorption in specific relationship to ions, soil particles, minerals, and plant roots
- Define and describe cations and anions including their behavior, influence, distribution, and importance in farming
- Define and describe acids and bases in relation to the soil and soil chemistry
- Read and analyze pH/Eh charts

LESSON CONTENT
- Lecture/Presentation/Video
- Reading/Listening/Watching
- Group Discussion
- Nature Immersion & Observation of Natural Systems
- Tactile Interaction
- Reflection
- Group Challenges & Activities

LESSON ACTIVITIES
- Read & Discuss *Regenerative Soil* sections: "The Mineral Components of Soil", p. 17–88

- Watch & Discuss *Regenerative Soil, the Online Course* sections: "The Mineral Components of Soil pt 1 & 2"
- Develop a model on paper, digitally, or physically to illustrate the role of photosynthesis and cellular respiration in the cycling of carbon among the biosphere, atmosphere, hydrosphere, and geosphere.
- Develop a quantitative model to describe the cycling of carbon among the hydrosphere, atmosphere, geosphere, and biosphere.
- *Test for Minerals* – students will test various soils for their soluble mineral content. The common water-based and colorful soil tests for gardeners can be used.
- *Test for REDOX* – students will test various soils for their Eh levels (may be only possible at a university with a platinum electrode!)
- *Test for pH* – students will test various soils for their pH levels.
- *Test for CEC* – students will test various soils for their CEC.
- *Test for Paramagnetism* – students will test various soils for their paramagnetic levels using a paramagnetic meter.
- Students will create a presentation of all the soils they tested using a variety of tests to create a description of what kind of soil it is and what the benefits or drawbacks are for that soil type.
- Students will create an essay, audio, video, or presentation to describe how Eh and pH are related but different
- Students will create an essay, audio, video, or presentation to describe how Eh and pH affect nutrient availability for plants
- Students will create an essay, audio, video, or presentation to describe how important Eh is to include in soil analysis and management

ASSESSMENTS
- Group Discussion (observed and through direct interaction)
- Group Activities, Artistic Representations, and Projects
- Written and Verbal Quizzes
- Identification Games/Exercises
- Student presenting (reteaching) content
- Student Notes
- Reflective Writing on New Learning

DIFFERENTIATIONS

Audiobook, Video-based instruction, teaching outside, having soil within reach for tactile interaction, with background music or in silence, discussion-based instruction, and learning 1:1, in small groups, or large groups.

5. THE BIOLOGICAL COMPONENTS OF SOIL

NGSS

LS2.A: Interdependent Relationships in Ecosystems
LS2.B: Cycles of Matter and Energy Transfer in Ecosystems
LS2.C: Ecosystem Dynamics, Functioning, and Resilience
ESS2.A: Earth Materials and Systems
ESS2.D: Weather and Climate
ESS2.E: Biogeology
ESS3.A: Natural Resources
LS1.C: Organization for Matter and Energy Flow in Organisms

NGSS

CONTENT STANDARD A – G

PES

2.6 Students are able to identify, describe, and present the different components of soil, the soil food web, photosynthesis in relation to the soil food web, and ways to improve soil and soil food web interactions

2.7 Students are able to identify, describe, illustrate, and present the various interactions
and function of trees and forests in relation ecosystems, natural cycles, precipitation, watersheds, soil, fungi, climate change, wildfires, and all biodiversity

RATIONALE

Students need to understand the biological components of the soil in order to have a holistic fluency to work with and analyze soil.

OBJECTIVES

Students will be able to…
- List and describe the essential members of the soil food web
- Describe and define the imperative roles and benefits of the soil food web for plants, in the soil, and in ecosystems
- Describe and define soil organic matter, its role, its states, and its benefits
- Describe and define aerobic, facultative, and anaerobic states especially in relation to the soil

- Describe and define the limitations of our understanding of the microbiological world of soil
- Describe and define how plants get their nutrition especially from the rhizosphere and atmosphere both in biological and ionic terms
- Describe and define the rhizophagy cycle and its implications
- Describe and define endophytes, their roles, and their benefits
- List, describe and define beneficial rhizobacteria, mycorrhizal fungi, and endophytes
- List and describe many composting, their benefits, and their uses
- Describe and define plant exudates, their function, their implications, and their benefits
- Describe and define soil food web products and their benefits
- Describe and define the 4 stages of plant health and what is needed to achieve each of these levels

LESSON CONTENT
- Lecture/Presentation/Video
- Reading/Listening/Watching
- Group Discussion
- Nature Immersion & Observation of Natural Systems
- Tactile Interaction
- Reflection
- Group Challenges & Activities

LESSON ACTIVITIES
- Read & Discuss *Regenerative Soil* sections: "The Biological Components of Soil ", p. 89–128 and "Humans & Soil", p. 137–143
- Watch & Discuss *Regenerative Soil, the Online Course* sections: "The Biological Components of Soil pt 1 & 2"
- Watch & Discuss *The Plant Health Pyramid*, the youtube video by John Kempf
- Watch & Discuss *Redox Potential (Eh) and pH as indicators of Soil, Plant and Animal Health and Quality.* A free Thinkific course hosted by Regen.ag Academy. https://academyregenag.thinkific.com/courses/redox-potential-eh-and-ph-as-indicators-of-soil-plant-and-animal-health-and-quality

- Watch & Discuss *Symphony of the Soil*, the documentary
- Watch & Discuss *Symphony of the Soil*, the documentary
- Watch & Discuss *Kiss the Ground*, the documentary
- Watch & Discuss the online courses by John Kempf: *Plant Health Pyramid* and *Precision Ag Plant Nutrition Management*. https://academyregenag.thinkific.com/courses/
- Watch & Discuss online Soil Food Web courses by Dr. Elaine Ingham
- Watch & Discuss *The Advanced Permaculture Student Online* program's "Soil" sections especially those featuring Dr. Elaine Ingham, Matt Powers, Cuauhtemoc Villa, and Michael Wittman: www.advancedpermaculturestudent.online
- *Test Your Soil For Biology* – students will gather samples of soil and either send them out to be tested by a soil food web lab or view them under a microscope in dilution to analyze them using *Regenerative Soil* p. 205 as a guide for ratios and numbers for healthy compost as a benchmark
- *Pitfall Trap Soil Testing* – students will test various soils using a pitfall trap to count the indigenous larger members of the soil food web – use *Regenerative Soil* p. 141 for directions.
- *Burlese Funnel Soil Testing* – students will test various soils using a burlese funnel method to count some smaller, though still visible, members of the soil food web – use *Regenerative Soil* p. 141 for directions.
- *Soil Testing by Hand for Life* – students will test various soils by digging through them with their hands in search of life – use *Regenerative Soil* p. 141 for directions.
- *BRIX testing* – if possible to acquire a BRIX meter, students will test the BRIX of plant sap from multiple plants over the course of a week during the same time of day (early morning). Students will analyze and discuss the results.
- *Plant Sap Analysis* – if possible, students will send in plant leaves to be tested. Students will analyze and discuss the results.
- *Compare Tests* – students will compare their biological test results with their soluble mineral, plant sap analysis, BRIX, pH, Eh, and paramagnetism test results and generate a comprehensive essay on the results with assertions as to the relative strengths and weaknesses of the tested soil given these test results

ASSESSMENTS
- Group Discussion (observed and through direct interaction)

- Group Activities, Artistic Representations, and Projects
- Written and Verbal Quizzes
- Identification Games/Exercises
- Student presenting (reteaching) content
- Student Notes
- Reflective Writing on New Learning

DIFFERENTIATIONS

Audiobook, Video-based instruction, teaching outside, having soil within reach for tactile interaction, with background music or in silence, discussion-based instruction, and learning 1:1, in small groups, or large groups.

6. THE ACTIONS

NGSS

LS2.A: Interdependent Relationships in Ecosystems
LS2.B: Cycles of Matter and Energy Transfer in Ecosystems
LS2.C: Ecosystem Dynamics, Functioning, and Resilience
PS3.D: Energy in Chemical Processes
ETS1.B: Developing Possible Solutions
ESS2.A: Earth Materials and Systems
ESS2.E: Biogeology
ESS3.A: Natural Resources

NGSS

CONTENT STANDARD A – G

PES

3.2 Students are able to create an advanced permaculture design for a home site – one that addresses water, waste, food, soil building, energy, and shelter
3.3 Students are able to design, set up, and manage a small outdoor garden, indoor garden, greenhouse garden, and larger outdoor garden
4.4 Students are able to make and apply thermophilic (hot) compost and vermicompost (using worms) to process food waste as well as compost teas and extracts
4.5 Students are able to use a microscope to identify soil food web organisms and determine soil, compost, compost extract, and compost tea quality

RATIONALE

Students need to know a diversity of methods, strategies, and techniques for managing soil and their individual effects to dynamically and with confidence manage and remediate soils.

OBJECTIVES

Students will be able to…
- Define, list, and describe cover crops and their benefits to soil
- Define, list, and describe biofertilizers and biostimulants
- Describe the attributes and the benefits of a great diversity of their own mycorrhizal, bacterial, and endophytic inoculants as well composts, specialized agars and broths, bioreactors, ferments, biostimulants, and biofertilizers

- Create of a great diversity of their own mycorrhizal, bacterial, and endophytic inoculants as well composts, compost teas, specialized agar, bioreactors, biochar, ferments, biostimulants, and biofertilizers
- Use a variety of regenerative amendments to improve soils including rock dusts
- Make a holistic grazing plan

LESSON CONTENT
- Lecture/Presentation/Video
- Reading/Listening/Watching
- Group Discussion
- Nature Immersion & Observation of Natural Systems
- Tactile Interaction
- Reflection
- Group Challenges & Activities

LESSON ACTIVITIES
- Read & Discuss *Regenerative Soil* sections: "The Actions", p. 144–208 and "Soil Remediation" p. 209–217
- Watch & Discuss *Regenerative Soil, the Online Course* sections: "The Actions pt 1 and 2"
- Watch & Discuss Chris Trump's Natural Farming videos on Youtube
- Watch & Discuss *The Advanced Permaculture Student Online* program's "Soil" sections especially the How To sections: www.advancedpermaculturestudent.online
- Read & Discuss BEST MANAGEMENT PRACTICES: JOHNSON-SU COMPOSTING BIOREACTORS by David Johnson and Patrick DeSimio: https://www.csuchico.edu/regenerativeagriculture/bioreactor/david-johnson.shtml
- *Build a Bioreactor* – students will build a bioreactor following BEST MANAGEMENT PRACTICES: JOHNSON-SU COMPOSTING BIOREACTORS by David Johnson and Patrick DeSimio
- *Create a Compost* – students will choose a compost method and create fully decomposed compost
- *Which Compost Is Best?* – students will develop several kinds of compost and compare them using soil testing and plant trail experimentation

- *Natural Farming Preps & Solutions* – students will create natural farming IMO preps and solutions for use in a garden or on a farm
- *Plant a Cover Crop* – students will plan and plant a cover crop for a garden, farm, or field taking soil tests before and after to measure and analyze the effects.
- *Mycorrhizal Inoculant Prep* – students will cultivate one or more mycorrhizal inoculants following the preps in *Regenerative Soil* (both ECM and AMF)
- *Bacterial Inoculant Prep* – students will cultivate one or more bacterial inoculants following the preps in *Regenerative Soil*
- *Endophytic Inoculant Prep* – students will cultivate one or more endophytic inoculants following the preps in *Regenerative Soil*
- *Trichoderma Inoculant Prep* – students will cultivate one or more trichoderma inoculants following the preps in *Regenerative Soil*
- *Make Biochar* – students will turn biomass into biochar using the conservation burn technique with supervision and adhering to local/state/federal guidelines for your area
- *Make Biofertilizer* – students will create a biofertilizer barrel system
- *Develop a Grazing Plan* – students will develop a grazing plan for a real-life site

ASSESSMENTS
- Group Discussion (observed and through direct interaction)
- Group Activities, Artistic Representations, and Projects
- Written and Verbal Quizzes
- Identification Games/Exercises
- Student presenting (reteaching) content
- Student Notes
- Reflective Writing on New Learning

DIFFERENTIATIONS
Audiobook, Video-based instruction, teaching outside, having soil within reach for tactile interaction, with background music or in silence, discussion-based instruction, and learning 1:1, in small groups, or large groups.

7. MAKING YOUR PLAN

NGSS

LS2.A: Interdependent Relationships in Ecosystems
LS2.B: Cycles of Matter and Energy Transfer in Ecosystems
LS2.C: Ecosystem Dynamics, Functioning, and Resilience
LS4.D: Biodiversity and Humans
PS3.D: Energy in Chemical Processes
ETS1.B: Developing Possible Solutions
ESS2.A: Earth Materials and Systems
ESS2.C: The Roles of Water in Earth's Surface Processes
ESS2.D: Weather and Climate
ESS2.E: Biogeology
HS-ESS3-2. Evaluate competing design solutions for developing, managing, and utilizing energy and mineral resources based on cost-benefit ratios.
HS-ESS3-4. Evaluate or refine a technological solution that reduces impacts of human activities on natural systems.
ESS3.A: Natural Resources
ETS1.B: Developing Possible Solutions

NSES

CONTENT STANDARD A – G

PES

3.2 Students are able to create an advanced permaculture design for a home site – one that addresses water, waste, food, soil building, energy, and shelter

3.3 Students are able to design, set up, and manage a small outdoor garden, indoor garden, greenhouse garden, and larger outdoor garden

4.4 Students are able to make and apply thermophilic (hot) compost and vermicompost (using worms) to process food waste as well as compost teas and extracts

4.5 Students are able to use a microscope to identify soil food web organisms and determine soil, compost, compost extract, and compost tea quality

4.14 Students have participated in regular acts of large-scale land restoration

RATIONALE

Students need to understand how to organize and prioritize different aspects of soil management to properly manage the hierarchy of function within soil. Students need to be able to develop a sophisticated soil management plan that will remediate and regeneratively enrich soil use overtime for agriculture and horticulture. Students need to be able to apply their understanding and education in soil management through making plans, implementing them, measuring the process and the outcomes, analyzing the data, and making informed new decisions as a soil manager.

OBJECTIVES

Students will be able to…

- List and describe the solutions to all of the common problems, deficiencies, toxicities, and antagonism found in the soil
- Organize solutions into a plan of action that saves money, time, and energy to quickly remediate or enrich soil
- Define and describe how soil coherence can be achieved

LESSON CONTENT

- Lecture/Presentation/Video
- Reading/Listening/Watching
- Group Discussion
- Nature Immersion & Observation of Natural Systems
- Tactile Interaction
- Reflection
- Group Challenges & Activities

LESSON ACTIVITIES

- Read & Discuss *Regenerative Soil* sections: "Make Your Plan ", p. 218–225
- Watch & Discuss *Regenerative Soil, the Online Course* sections: "Make Your Plan"
- *Make Your Plan* – students will copy on paper or digitally recreate the charts in "Make Your Plan" and fill them out with the outline of their plan, and then students will write out a rationale for their plan explaining why they chose the timing, inputs, methods, and tests in their particular management plan
- *Present Your Plan* – students will present their plan in an audio/video/slideshow/presentation format to peers, teachers, mentors, and even social media friends if one is comfortable to get feedback
- *Enact Your Plan* – students will test out their soil management plans with an actual piece of ground for a season tracking the before and after status of the soil, analyzing it, reflecting on the results, and then writing up a review of the entire process and what one would do differently next season.

ASSESSMENTS.

- Group Discussion (observed and through direct interaction)
- Group Activities, Artistic Representations, and Projects
- Written and Verbal Quizzes
- Identification Games/Exercises
- Student presenting (reteaching) content
- Student Notes
- Reflective Writing on New Learning

DIFFERENTIATIONS

Audiobook, Video-based instruction, teaching outside, having soil within reach for tactile interaction, with background music or in silence, discussion-based instruction, and learning 1:1, in small groups, or large groups.

PROJECTS

Project-based assessments allow students to demonstrate their understanding authentically. They value the experience, and for teachers, it's best practice and shows us how effective our teaching has been: it's where the real learning gets exposed.

Regenerative Soil Project Proposal Samples

These are sample projects to show you what's possible - don't hesitate to adapt them, and you can use them in your program as is as well. They are designed as summative, performance-based assessments of learning, but they are also open-ended, generalized, and have variation and option built into them. I highly recommend carrying that offering of choice forward on to the student as student-centered programs drive greater engagement and deeper learning.

Guiding Questions
- *What's your regenerative soil project?*
- *What are your goals with this project?*
- *What tests will you rely upon?*
- *What are the metrics for success?*
- *What is the timeline for this project?*
- *How is this soil project regenerative?*

SAMPLE #1
BIOFERTILIZER CULTIVATION

Project
Students will cultivate a beneficial bacteria or fungi from The Actions section of *Regenerative Soil*.

Goals
To learn how to cultivate high quality biofertilizers at home.

Tests
- *A/B Comparison with Plants* - for example: testing the homegrown biofertilizer vs the commercial biofertilizers with a no-biofertilizer control. Size, yield, flavor, BRIX, weight, # of leaves, etc. can be compared.
- *A/B Comparison with Microscopes* - for example: testing the numeracy of the homegrown vs the commercial biofertilizer's microbial population per mL or mg with a control sample from a natural (un-amended) local soil site.

Metrics for Success
- Healthy Plant Response
- Equivalency or Superiority to Commercially-Available Biofertilizers

Timeline
Weeks - Months (depending on the microbe being cultivated and test conducted)

Rationale
Knowing how to cultivate your own biofertilizing bacteria and fungi allows you to support plants at the cellular level in a way that traditional fertilizers and purely mineral amendments cannot even hope to mimic. Not even compost can compete if it lacks the necessary microbes within it - these are microbes that have special

relationships with plants that are not ever present in all systems, so adding them is critical and knowing how to do it from home can save money and generate a secondary income if one chooses to scale up their cultivation. It's a powerful set of skills that have real economic demand and power in agricultural settings and beyond.

Examples

Students may cultivate arbuscular mycorrhizal fungi (AMF), ectomycorrhizal fungi (ECM), *Trichoderma*, Rhizobium, Azospirillum, Azotobacter Chroococcum, B. megaterium, Dark Septate Endophytes, and IMOs of all sorts following the guides and recipes in The Actions section of *Regenerative Soil*.

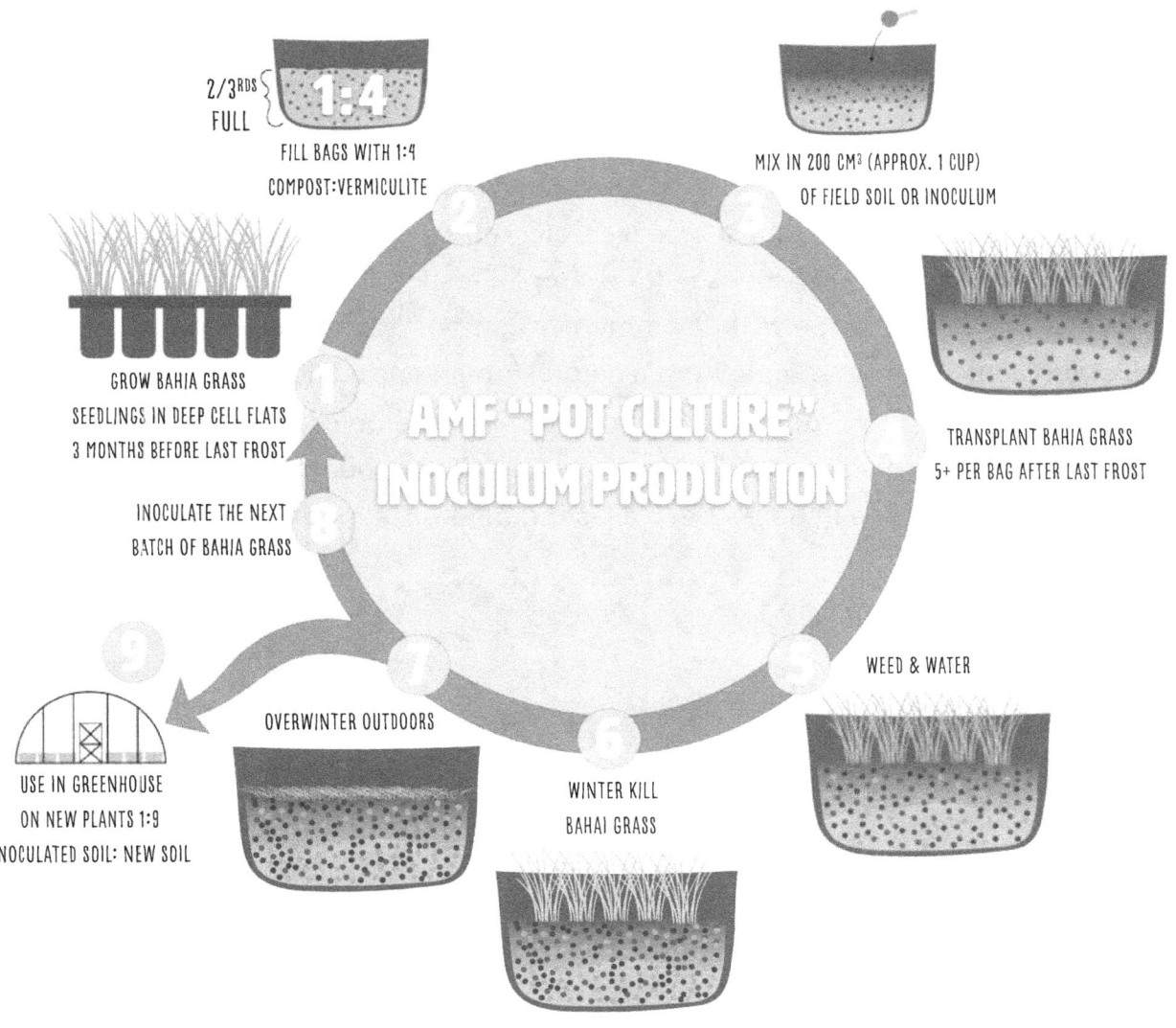

SAMPLE #2
COMPOST CULTIVATION

Project
Students will build and manage a compost pile to maturation and apply it to test its effects.

Goals
To learn how to make high quality compost and to recognize its efficacy.

Tests
- *A/B Comparison with Plants* - for example: testing the homemade compost vs the commercially acquired compost on plants with a control plant without compost. Size, yield, flavor, BRIX, weight, # of leaves, etc. can be compared.
- *A/B Comparison with Microscopes* - for example: testing the numeracy and diversity of the homemade vs the commercial compost's microbial population per mL or mg with a control sample from a natural (un-amended) local soil site.
- *Before & After Soil C & Biodiversity Testing* - testing soils before application of compost for biodiversity and soil organic matter (C) and then testing soils a year after application to see if there's been a lasting difference made by that application the year prior.

Metrics for Success
- Healthy Plant Response
- Higher Soil Organic Matter Levels
- Superiority to Commercially-Available Compost in Effect & Biodiversity
- Soils Retain More Biodiversity

Timeline
Weeks - Months (depending on the microbe being cultivated and test conducted)

Rationale

Knowing how to cultivate your own compost for adding soil organic matter to the soil is important not just for the soil but for responsible and ethical handling of our organic waste products.

Examples

Students may create a Johnson-Su bioreactor compost, a thermophilic compost, a mouldering compost, a vermicompost system, a semi-aerated compost, an EM compost, a woody Jean Pain compost, or an IMO-3 or IMO-5 prep.

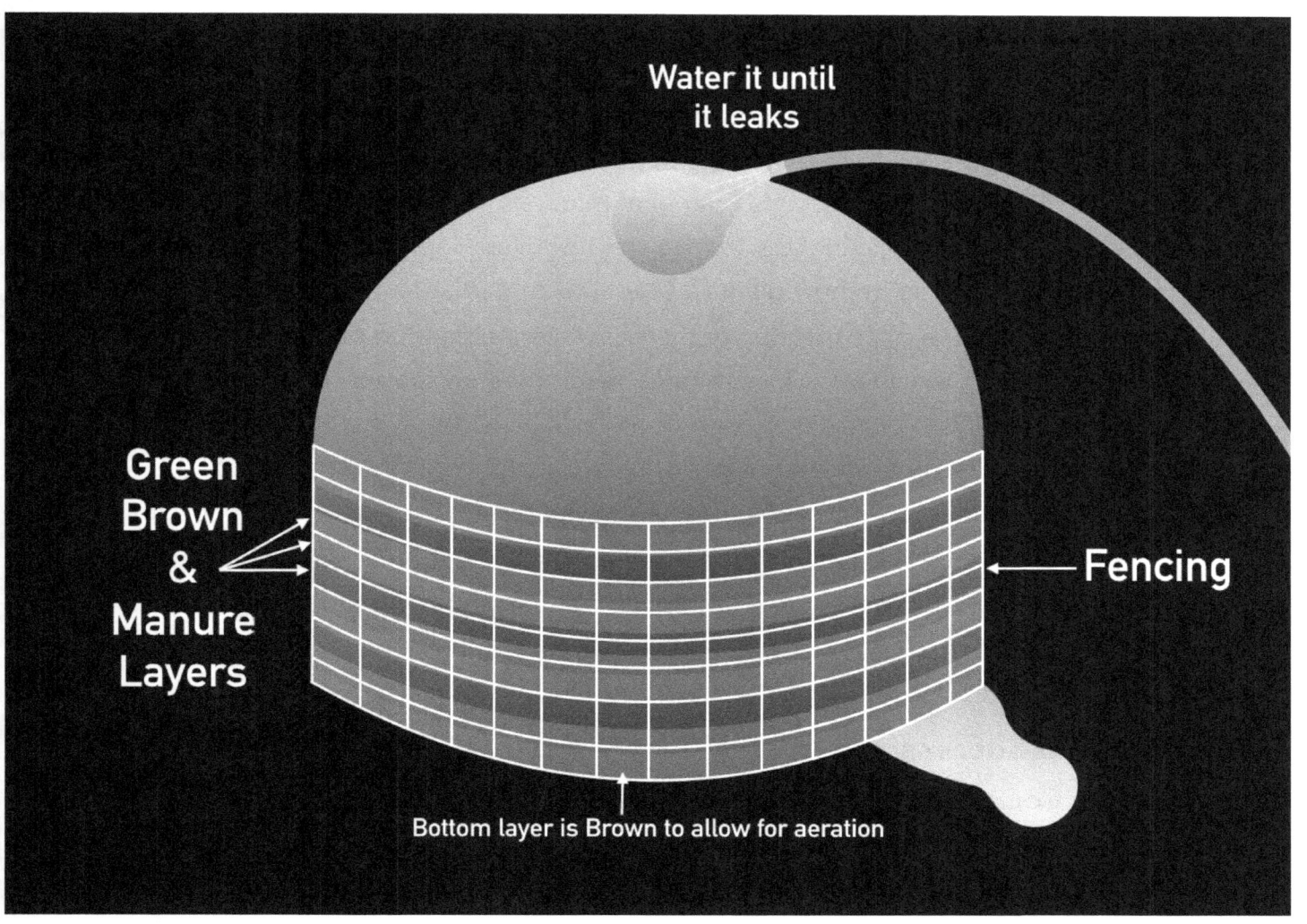

A Thermophilic Compost Setup

SAMPLE #3
SOIL REMEDIATION

Project
Students will remediate a contaminated soil using techniques found in the Soil Remediation section of *Regenerative Soil*.

Goals
To learn how to diagnose and remediate contaminated and toxic soils by actually doing it.

Tests
- *Before & After Soil Testing* - testing soils before remediation and then testing soils a year after remediation to see if there's been a lasting difference made by our efforts. Individual minerals or a full panel of minerals can be tested.
- *Before & After Plant Sap Analysis Testing* - testing plant sap before remediation and then testing plant sap on new plants a year after remediation to see if there's been a change in the rate of absorption and types of contamination present.
- *A/B Comparison with Plants* - for example: testing the untreated soil vs the remediated soil with plants in a controlled trial like inside a greenhouse, in pots, or in the field if some areas have not been remediated.
- *A/B Comparison with Microscopes* - for example: testing the numeracy, ratios, and diversity of the untreated soil vs the remediated soil's microbial populations per mL or mg with a control sample for comparison from a natural (un-amended) local soil site.

Metrics for Success
- Toxin or Contamination Reduction to Safe Levels
- Healthy Plants with Low or No Contamination or Toxins Present
- Healthy Soil Microbes Present

Timeline
Months - Year

Rationale
Knowing how to diagnose and remediate soil is an incredibly critical skill in areas with soil contamination (especially since these areas are becoming increasingly common and increasingly in overlap with our homes, communities, farmland, coastlines, riparian zones, and recreational areas.) We need to address the elephant in the room that is widespread pollution and contamination of our soils, water, and air.

Examples
Students may remediate soils using any of these methods: phytoextraction, phytodegradation, phytovolitization, rhizodegradation, rhizofiltration, phytostabilization, microbial remediation, chemical remediation, mechanical/physical remediation, technological remediation.

SAMPLE #4
NATURAL FARMING IMO PREPS

Project
Students will know how to create natural farming IMO preps and use them in a garden, farm, orchard, or landscaping setting.

Goals
To learn how create high quality IMO preps and recognize their efficacy.

Tests
- *A/B Comparison with Plants* - for example: testing the IMO vs hot compost with a no-input control. Size, yield, flavor, BRIX, weight, # of leaves, etc. can be compared.
- *A/B Comparison with Microscopes* - for example: testing the numeracy of the IMO vs hot compost's microbial population per mL or mg with a control sample from a natural (un-amended) local soil site.
- *Before & After Soil Testing* - testing soils before IMO additions and then testing soils a year after IMO's were added to see if there's been a lasting difference made by our efforts. Individual minerals, microbes, SOM, or a full panel of minerals can be tested.
- *Before & After Plant Sap Analysis Testing* - testing plant sap before IMO additions and then testing plant sap on the same plants 1 week, 2 weeks, 3 weeks, and/or at harvest with a control plant that was not given IMOs for comparison.

Metrics for Success
- Healthy Plant Response
- Increased Soil Biology & Soil Organic Matter
- Equivalency or Superiority to Hot Compost

Timeline
Weeks - Full Year

Rationale
Knowing how to cultivate our own indigenous microorganisms is vital for all farmers and gardeners alike because IMOs are always superior to imported foreign and commercially cultivated microbes. Knowing what the effect are of these inputs in our area and context is also critical to know how to apply them strategically and beneficially.

WATER SOLUBLE CALCIUM

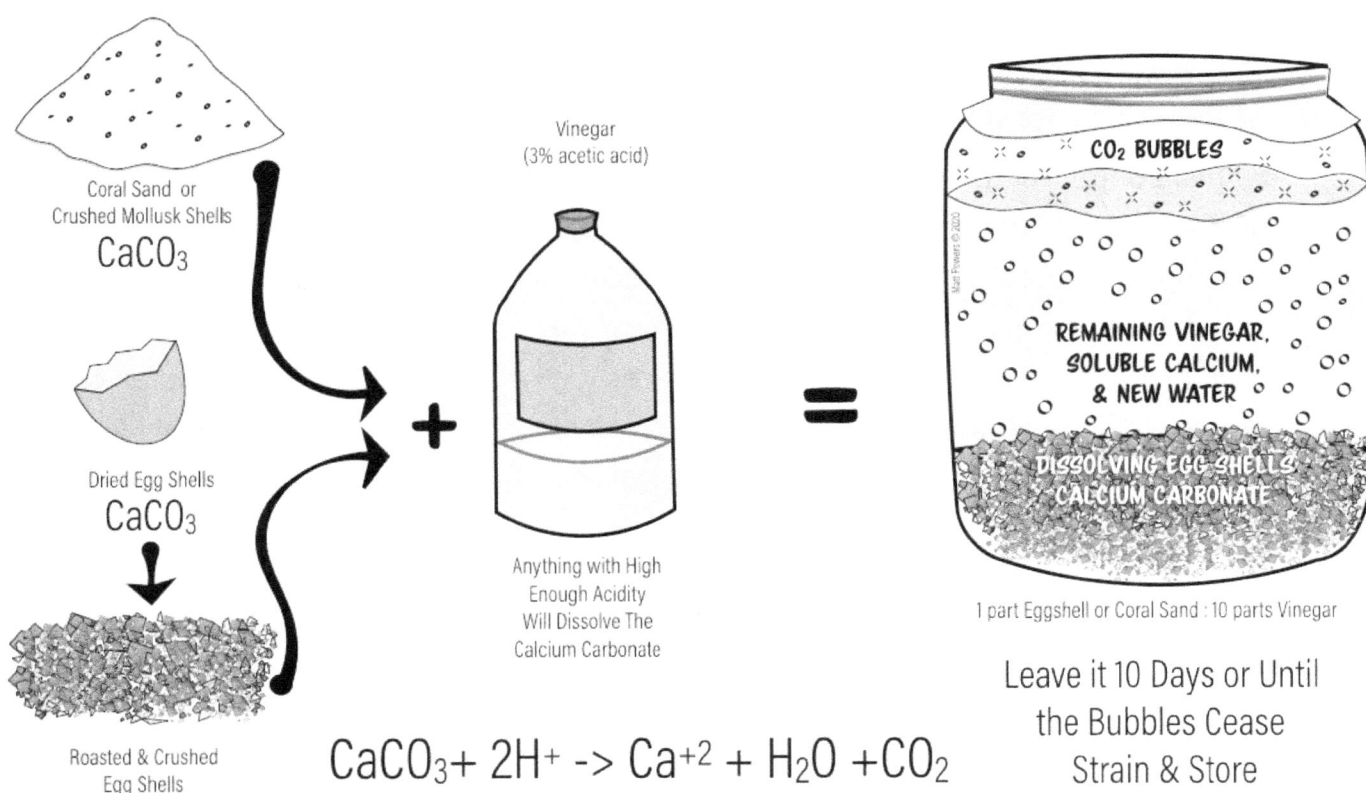

$$CaCO_3 + 2H^+ \rightarrow Ca^{+2} + H_2O + CO_2$$

1 part Eggshell or Coral Sand : 10 parts Vinegar

Leave it 10 Days or Until the Bubbles Cease
Strain & Store

SAMPLE #5
COMPOST TEA BREWING

Project
Students will make compost tea and test its efficacy on plants and soils.

Goals
To learn how to brew effective compost tea and recognize its utility and effect.

Tests
- *A/B Comparison with Plants* - for example: testing the home-brewed compost tea vs the commercially available compost tea with a no-compost tea control. Size, yield, flavor, BRIX, weight, # of leaves, etc. can be compared.
- *A/B Comparison with Microscopes* - for example: testing the numeracy of the home-brewed vs the commercial compost tea's microbial population per mL or mg with a control sample from a natural (un-amended) local soil site.
- *Before & After Soil Testing* - testing soils before compost tea additions and then testing soils a year or a season after compost teas were added to see if there's been a lasting difference made by our efforts. Microbes or SOM can be tested.
- *Before & After Plant Sap Analysis Testing* - testing plant sap before compost tea was added and then testing plant sap on the same plants 1 week, 2 weeks, 3 weeks, and/or at harvest with a control plant that was not given compost tea for comparison.

Metrics for Success
- Healthy Plant Response
- Increase in Soil Biology
- Increase in SOM
- Equivalency or Superiority to Commercially-Available Compost Teas

Timeline
Weeks - Months (depending on the microbe being cultivated and test conducted)

Rationale
Knowing how to cultivate your ow

SAMPLE #6
FERMENTATION "BIOFERTILIZER" CREATION

Project
Students will create an anaerobic fermentation bioreactor for "biofertilizer" (which is a colloquial term for an anaerobic fermentation for SOM, nutrients, and minerals and not to be confused with microbes that are termed biofertilizer industry-wide and in scientific literature.) They will create it, manage it, use it, and then test its efficacy in comparison to other natural fertilizers.

Goals
To learn how to effectively use anaerobic fermentation to create a powerful "biofertilizer" soil amendment.

Tests
- *A/B Comparison with Plants* - for example: testing the anaerobic "biofertilizer" vs the hot compost with a no-input control. Size, yield, flavor, BRIX, weight, # of leaves, etc. can be compared. Hot compost could be substituted or be one of a panel of compost methods and amendment types compared.
- *A/B Comparison with Microscopes* - for example: testing the numeracy, ratios, and diversity of the anaerobic "biofertilizer" vs the hot compost with a control sample from a natural (un-amended) local soil site. Hot compost could be substituted or be one of a panel of compost methods and amendment types compared.
- *Before & After Soil Testing* - testing soils before biofertilizer additions and then testing soils a year or a season after biofertilizers were added to see if there's been a lasting difference made by our efforts. Microbes, minerals, and/or SOM can be tested.
- *Before & After Plant Sap Analysis Testing* - testing plant sap before biofertilizers was added and then testing plant sap on the same plants 1 week, 2 weeks, 3 weeks, and/or at harvest with a control plant that was not given biofertilizer for comparison.

Metrics for Success
- Healthy Plant Response
- Equivalency or Superiority to Hot Compost, Compost Tea, or IMO-3
- Increased SOM, beneficial microbial populations, and bioavailable minerals

Timeline
Months - Year

Rationale
Knowing how to cultivate your own anaerobic fermentation safely and for an effective beneficial result is valuable for gardeners, farmers, and anyone who works with soil or grows food.

SAMPLE #7
A/B TESTING NATURAL INPUTS

Project
Students will test and compare and contrast 2 or more natural inputs that can be made in their bioregion. This could be Johnson-Su compost vs mouldering, IMO-3 vs hot compost, biofertilizer inoculants vs compost tea soil soaks, etc. but along a certain focused set of tests: leaf size, leaf #, # of fruits, yield weight, plant weight, plant height, plant width, BRIX of fruit, etc. The comparison depends on what the student is interested in, what they are capable of doing, and what is locally available.

Goals
To test and see what natural input methods work best in a given bioregion or site.

Tests
- *A/B Comparison with Plants* - for example: testing the anaerobic "biofertilizer" vs the hot compost with a no-input control. Size, yield, flavor, BRIX, weight, # of leaves, etc. can be compared.
- *A/B Comparison with Microscopes* - for example: testing the numeracy, ratios, and diversity of the IMO-5 vs the hot compost vs mouldering (or any other inputs) with a control sample from a natural (un-amended) local soil site.
- *Before & After Soil Testing* - testing soils before a series of natural input additions and then testing soils a year or a season after the inputs were added to see if there's been a lasting difference made by our efforts. Microbes, minerals, and/or SOM can be tested.
- *Before & After Plant Sap Analysis Testing* - testing plant sap before a series of natural input additions were added and then testing plant sap on the same plants 1 week, 2 weeks, 3 weeks, and/or at harvest with a control plant that was not given compost tea for comparison.

Metrics for Success
- Students will be able to identify a hierarchy of efficacy among natural inputs for your bioregion
- Students will be able to identify and differentiate the benefits of different inputs

Timeline
Weeks - Year (depending on the microbe being cultivated and test conducted)

Rationale
Knowing what works best for our soils and bioregion is critical - we can't assume that the natural inputs in our area with their unique collection of nutrients, microbes, and minerals will behave in exactly the same way as other bioregions. Bokashi may be better than hot compost for example. This evaluation of inputs gives students an insight into what will be ideal for them to use in gardens, on farms, in orchards, and in landscaping.

FINAL THOUGHTS

Soil is more critical than we can convey, understand, or measure. We only have identified 1% of soil microbes – we are at the beginning, but it's an excited new frontier of innovation, exploration, and experimentation with seriously amazing rewards for everyone and the planet in the mix. I hope you can share this message with as many as possible!!

Thank you again for sharing regenerative soil with others! Without your help, this knowledge and new pathway wouldn't be spreading as fast as it is! You're helping others, helping the earth, and helping the future!! THANK YOU!!

Use this book along with *Regenerative Soil*, the book and the online course at
www.RegenerativeSoilScience.com,
and start getting your hands in the soil today!!

Grow Abundantly, Learn Daily, & Live Regeneratively,

Matt Powers

ABOUT THE AUTHOR

Matt Powers is an author, educator, seed saver, gardener, and entrepreneur focused on radically transforming the entire K-12 education system through the collegiate system as well as the economy such that it aligns with regenerative science, natural principles, and permaculture ethics: Earth Care, People Care, & Future Care. Matt, a former public high school teacher with a Masters degree in Education, is the author of the first government accredited permaculture curriculum in North America (fully cited, peer-reviewed, & aligned to national standards), and his work continues to spread in schools, colleges, and universities globally with over a dozen books in 6 languages and 9 online courses.

MATT IS ON A MISSION TO EMPOWER PEOPLE EVERYWHERE TO LIVE MORE REGENERATIVELY.

SEE WHAT STUDENTS ARE SAYING...

"It is my opinion that it is **the best permaculture course in the world**"
  -James Webb

"Matt Powers - you are the true reflection of what permaculture means with caring for people not just the paperwork. **This is my destiny and I knew as soon as I signed up that this was the course, that it would change my life.**"
  - Beth Healey

"This course is World Class! I love the info. **This is the best course I have ever taken on this subject. Matt Powers ROCKS!!!**"
  -Mike Garcia

"The PDC from a few years ago truly changed my way of looking at everything. Your course has up-scaled the learning to **reach deeper** into the way I view the interior and exterior worlds...**micro to macro**."
  -Jennifer Brennan

"I appreciate your **tremendous, inspiring work** and this exciting learning experience so much."
  - Yoko Fujimoto

"This course is fantastic Matt! Your energy amps up the lessons and I love the audiobook format. I'm honored to be a part of this. Thank you brother."
  -Nik Cudnik

"I am an environmental science student at Georgia State and **I have learned more in this course than my 4 years at this University.**"
  - Alex Kerr

"Geoff Lawton's PDC completely changed my understanding of nature, but this course is just taking it to a whole new level. **Future now looks more feasible.**"
  - Thiago

"This course has been **the most valuable course I've ever taken**, and I cant wait to go back through it all a second and third time. This is **a true gift to the world** Mr. Powers. Thank you!"
  -Mark

GET YOUR DUAL PERMACULTURE DESIGN CERTIFICATION WITH OVER 70 EXPERTS FROM AROUND THE WORLD SPANNING DOZENS OF REGENERATIVE CAREER PATHS, LIFESTYLES, TECHNIQUES, PERSPECTIVES, & SOLUTIONS

THE ADVANCED PERMACULTURE STUDENT ONLINE

GET INSPIRED. GET EMPOWERED.

This is the **only Advanced Permaculture Certification course** based on a current, peer-reviewed, & fully cited advanced permaculture curriculum & taught by experienced professionals from a host of different regenerative career paths & led by author, gardener, life coach, & educator Matt Powers. Learn to blend regenerative disciplines inside & outside the garden & homestead scope of permaculture design & into a broader application of permaculture into everyday life, design, business, architecture, food, energy, education, & even governance.

www.AdvancedPermacultureStudent.com

REGENERATIVE SOIL
THE 12-WEEK ONLINE COURSE WITH MATT POWERS

GET YOUR REGENERATIVE SOIL CERTIFICATION AND LEARN THE SCIENCE, TECHNIQUES, & SOLUTIONS WITH VIDEO & AUDIO

GET YOUR QUESTIONS ANSWERED

Grow Amazing Food

Build Rich & Regenerative Soil

Remediate & Restore Damaged & Toxic Soils

INCLUDES INCREDIBLE BONUSES:

— **Regenerative Entrepreneurs & Experts**

The 9-week Course

— **Transformative Ebooks:**

Regenerative Soil and The Permaculture Student 2 Workbook and Textbook

— **3 months of LIVE ZOOM Training w/Q&A**

with Matt Powers

WWW.REGENERATIVESOILSCIENCE.COM